TOO CUTE! BABY ANIMALS

BABY ARCTIC ANIMALS

RACHAEL MORLOCK

PowerKiDS press

Published in 2025 by The Rosen Publishing Group, Inc.
2544 Clinton Street, Buffalo, NY 14224

First Edition

Editor: Rachael Morlock
Book Design: Michael Flynn

Photo Credits: Cover, p. 1 evaurban/Shutterstock.com; (series background) Viktoriia LiSa/Shutterstock.com; p. 5 Don Landwehrle/Shutterstock.com; p. 7 Frank Fichtmueller/Shutterstock.com; p. 9 Reed Conrad/Shutterstock.com; p. 11 Giedriius/Shutterstock.com; p. 13 AndreAnita/Shutterstock.com; p. 15 CampCrazy Photography/Shutterstock.com; p. 17 Vladimir Melnik/Shutterstock.com; p. 19 Kricha/Shutterstock.com; p. 21 Wirestock Creators/Shutterstock.com.

Library of Congress Cataloging-in-Publication Data

Names: Morlock, Rachael, author.
Title: Baby Arctic animals / Rachael Morlock.
Description: [Buffalo] : PowerKids Press, [2025] | Series: Too cute! Baby animals | Includes bibliographical references and index.
Identifiers: LCCN 2024001833 (print) | LCCN 2024001834 (ebook) | ISBN 9781499445114 (library binding) | ISBN 9781499445107 (paperback) | ISBN 9781499445121 (ebook)
Subjects: LCSH: Animals–Arctic regions–Juvenile literature. | Animals–Infancy–Arctic regions–Juvenile literature. | Habitat (Ecology)–Arctic regions–Juvenile literature.
Classification: LCC QL105 .M67 2025 (print) | LCC QL105 (ebook) | DDC 591.3/92–dc23/eng/20240215
LC record available at https://lccn.loc.gov/2024001833
LC ebook record available at https://lccn.loc.gov/2024001834

Manufactured in the United States of America

CONTENTS

In the Arctic. 4
Arctic Animals. 6
Caribou Calves 8
Fox Kits 10
Bear Cubs 12
Seal Pups 14
Baby Belugas 16
Arctic Chicks. 18
Growing Up in the Arctic . . . 20
Glossary 22
For More Information 23
Index 24

In the Arctic

The Arctic is one of the coldest areas on Earth. It's an area around the North Pole that includes the Arctic Ocean and parts of Alaska, Canada, Greenland, Scandinavia, and Russia. **Permafrost** covers much of the land. In winter, chunks of ice come together and cover the water.

Arctic Animals

Long, dark winters make life in the Arctic difficult. That doesn't stop some creatures from living there all year long! Many Arctic animals are **mammals** that live on land or in the sea. As babies, they learn to survive in the cold Arctic **climate**.

musk ox

Caribou Calves

Caribou, also known as reindeer, are born in the Arctic **tundra** in June. A calf drinks milk and later shares plants with its mother to build its strength. Before winter, caribou leave the tundra for a long **migration** into forests where winter food can be found.

Fox Kits

Litters of Arctic fox kits are born in dens in the spring. Kits learn to hunt from their parents. After about six months, they can live on their own. Their short brown fur becomes a thick white or blue-gray coat in winter. It blends in with the snow!

Bear Cubs

Usually, two polar bear cubs are born together in the winter. For the first few months, they stay safe and warm in a snow den with their mother. When they are bigger, they learn to swim and hunt seals on the ice and in the water.

Seal Pups

Harp seal pups are born on the ice. Their soft white fur is **camouflage** that also keeps them warm. After about two weeks, dark hair grows beneath the white fur and turns their coat gray. As they grow, they spend more time in the water.

Baby Belugas

The Arctic Ocean is home to many whales, including belugas and narwhals. Baby belugas are born gray or brown and become white as they age. A thick layer of fat called blubber keeps them warm in icy waters. They migrate south in winter before the sea freezes over.

Arctic Chicks

Not all of the animals in the Arctic are mammals. Many are birds that come in the summer to make nests and lay eggs. The Arctic tern is one bird that **hatches** in the summer. In the winter, it flies all the way to **Antarctica**!

Growing Up in the Arctic

Many baby Arctic animals have to grow strong quickly to be ready for winter migrations. Others, like the snowy owl, live in the Arctic year-round. Whether it's winter or summer, these cute animals count on warm fur, cozy dens, and special skills for life in the Arctic.

GLOSSARY

Antarctica: The continent around the South Pole.

camouflage: Colors or shapes on animals that allow them to blend in with their surroundings.

climate: The average weather conditions of a place over a period of years.

hatch: To break open or come out of an egg.

mammal: Any warm-blooded animal whose babies drink milk and whose body is covered with hair or fur.

migration: The act of an animal or people moving from one location to another.

permafrost: A soil layer that is always frozen in very cold regions of the world, such as the Arctic.

tundra: Cold northern lands that lack forests and have permanently frozen soil below the surface.

FOR MORE INFORMATION

BOOKS

Hansen, Grace. *Arctic Hare.* Minneapolis, MN: Abdo Kids, 2020.

Hansen, Grace. *Narwhal.* Minneapolis, MN: Abdo Kids, 2020.

Sanderson, Whitney. *Meet a Baby Polar Bear.* Minneapolis, MN: Lerner Publications, 2024.

WEBSITES

Arctic Fox
kids.nationalgeographic.com/animals/mammals/facts/arctic-fox
Read more about the Arctic fox and its special skills for surviving the Arctic winter.

Polar Bear Facts
polarbearsinternational.org/education-center/just-for-kids/
Learn about the amazing polar bear, and find out what you can do to protect its Arctic home.

INDEX

B
birds, 18

C
camouflage, 14
caribou, 8
climate, 6

F
foxes, 10

I
ice, 4, 12, 14

M
mammals, 6, 18
migration, 8, 16, 20

N
North Pole, 4

P
polar bears, 12

S
seals, 12, 14

T
tundra, 8

W
whales, 16